GLOW

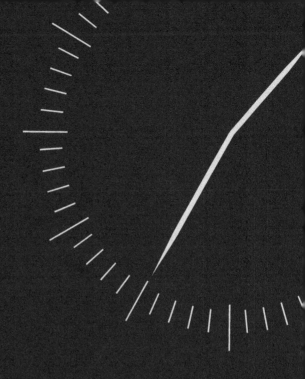

GLOW

Ann Hudson

Published by
Next Page Press
San Antonio, Texas
www.nextpage-press.com
© 2021 Ann Hudson. All rights reserved.

ISBN: 978-1-7366721-0-5

Cover photo by Allen Rein
Book design by Amber Morena

CONTENTS

GLOW

Afterglow

Radium Dial stood empty, fenced off,
abuzz. Boards on the windows, scrub weeds

pocking the dusty yard. The story quieted down.
After a while no one gazed through

the chainlink fence. When at last the factory
was leveled, locals picked over the rubble,

helping themselves to good bricks.
Schools were glad to reclaim the oak desks.

Even now radioactive fingerprints tarnish
everything: houses, a car dealership, a running track,

a lot that hosts a flea market on the weekends.
St. Columba Cemetery radiates.

We think what we can't see can't hurt us.
We think this deep in our bones.

Paradise

The day before he died, my father whispered
they were building him a new lab—

bigger than he really needed. *It's an honor,*
he confided, his hands fretting the sheets.

A cup of sponges to wet his mouth
stood on the nightstand. We sat quietly together,

imagining a spacious workspace: white, gleaming,
with glass doors shushing open, shushing closed.

I stood to carry another load of laundry
to the basement, where the chain tinked

against the bare bulb, and crickets wheezed
behind the hot water tank. After their modest wedding,

Marie and Pierre Curie bicycled across Brittany;
because her simple black dress would have tangled

in the spokes, she wore knickers and knee-stockings
and a straw boater with a narrow brim.

Back in Paris they returned to their workspace
behind the School of Physics: an abandoned shed

with a leaky roof, dirt floor, and a rusty cast-iron stove.
The light was dim. Specks of coal dust

ruined days of painstaking labor. It was blazing
in summer, freezing in winter, but the place

was so shabby no one thought to stop by
unannounced, and when the work was going well

the hours flew by. It was here they detected
two new elements, one of which she named

polonium for her home country, the other *radium*
for the energy it emitted seemingly without end.

By the time I returned to my father's bedroom
he was asleep, dreaming of his paradise.

Electric Fairy

I suppose I am the only person who is known as a dancer but who has a personal preference for Science.
—LOÏE FULLER

How many bolts of sheer, white silk were stitched
together to make the dancing dress of Loïe Fuller,

sensation of the Folies Bergère, darling of Parisian
high society, an American girl born in the outskirts

of what was barely Chicago, barely a muddy crossroads?
There she was, in the music hall famous for spectacle

and costume, gripping the bamboo sticks slipped
into pockets of her voluminous gown so she could twirl

her dress into a whirlwind of fabric, stagehands training
the heavy lights on her, operating the projectors,

glass plates, colored gelatins. She danced on stage alone
for forty-five minutes, lights searing her eyes,

the audience entirely silent, then erupting in applause.
It wasn't so much dance she was after as a way

to understand radiance. She'd written the Curies to ask
how she might fashion a garment that would generate

its own light, and they'd generously written back
a lengthy but discouraging response. Radium was

impractical, too expensive. Soon she was knocking
on their door on Boulevard Kellermann, accompanied

by a pack of technicians and stage hands, begging the Curies
to vanish to their lab for the day so she could set up

the required environment for the radium dance she'd
imagined, choreographed, and produced. For hours

she hung heavy drapes, arranged rugs, and projected
colored lights she'd designed herself. Eventually

Fuller would take out over a dozen patents for her stage lights,
her sets, her dresses, and her dances; the Curies

never patented their method for rendering radium
nor its medical applications, believing radium

for the good of humankind. How baffling to see light
as constructed and discovered, as proprietary

and universal, as illusion and as vision. How light whirls
and spins even as it seems to stay right where it is.

Marie Curie Addresses Pierre in Her Mind While She Inspects the Stove In the Vestibule

1906

I rush home from the train each night,
the girls asleep, twisted in their bedsheets.

The house has already cooled. Your father,
awake in the dark, leaves me to myself.

Into the stove I layer crumpled newspapers,
notes, condolence cards the world insists

on sending. Then kindling the children gather:
linden, honey locust, chestnut. I snap twigs,

bend the greens, and add the hard coal last:
six shovelfuls, one for every letter of your name.

Why Grow Old?

from early 20th century ad copy

Ladies, give your skin that radiant glow
with our face powder. Rinse with radium water

after a luxurious soak in radium mud.
Treat yourself to a nibble of radium

chocolate. Soothe your skin with radium cream.
Gentlemen, before you step out of the house

adjust your necktie that glows
in the dark, slip this luminescent watch

beneath your cuff link. Your children
will thrive on sandwiches made of radium bread

and radium butter. And before bed
don't forget to polish with radium toothpaste.

Try our radium heating pads. Insert
our radium suppositories to revive

what droops or sags! Tuck this radioactive
card beneath your scrotum!

Each morning sip this refreshing radium
tonic! We call it *perpetual sunshine.*

Petites Curies

It's true, the radium burned her fingertips,
left scars and numbness. But it

was worth the work. It was worth
the three years of strenuous labor

rendering seven tons of pitchblende
to yield a single gram of radium.

The money from the Nobel Prize
was only worth the laboratory it could build.

The work meant everything.
And once the War began, all research

stopped. What if she hadn't devised
and driven retrofitted military trucks

to bring radiology to wounded soldiers
at the front? Imagine her hair swept up

off her brow, her worn, black dress
oily after she changed a tire, or cleaned

a dirty carburetor, the arm band of the Red Cross
on her sleeve. She taught herself

anatomy. She taught herself to drive.
The wounded were brought one by one

to the darkroom, where Curie would conduct
the x-ray, showing the bones and organs,

showing the fragment of shell.
Sometimes the surgeon would operate

on the spot. For four years she drove
all over France, installing two hundred

radiological rooms, equipping twenty
motor cars she begged from patrons

and transforming them into mobile x-ray units
nicknamed *petites Curies*. And what's more

she figured out a way to siphon off
the gas that radium steadily emits,

capture the radon in thin, glass tubes,
and transport the delicate tubes

to the operating fields where surgeons
could inject the radon into injured tissue.

Radium was beautiful and it was good.
That is the only reckoning there was.

Work

1922

Our soldiers told time in the watery trenches,
and look: now every gentleman

with a watch of ours can see the numbers
in the evening light as he walks home.

Every mother rising in the dark
to soothe her child can know the hour.

We sweep the hands with paint that glows.
Just like we were taught we press the brushtips

between our lips. We bring them
to a shining point with our mouths, our glowing ohs.

Work

1923

This was where any young woman who could paint precisely
and in good time could make good money

and good money had been hard to come by
for a long time. Radium Dial had moved in

to the old high school, refitted the student desks
with lamps and a side caddy for supplies.

Our supervisors taught us lip-pointing. The paint
would give us a healthy radiance, bring color

to our cheeks, give us energy. A hundred of us
sat in rows, painted tiny numerals on watch dials

and made them shine. On breaks we daubed our skin,
our lips, our nails, and laughed to see ourselves aglow.

It was harmless fun. But we were the clocks,
the radium ticking on and on and on.

Work

1934

Girls were getting sick. A few at first,
then more. One woman stumbled

on the dance floor and her broken leg
wouldn't mend. Another had a toothache;

the dentist tugged on a single tooth
and her jaw crumbled. Some girls

were too weak to work again. Some died.
Company doctors stated their cause of death

was syphilis, and downplayed it
in the papers. But by the time

the five girls sued, one so skeletal
she had to be carried into court,

Radium Dial closed down. Six weeks later
and about six blocks away, we reported

back to work. New factory, new name:
Luminous Processes. Everything else

the same. Except they told us
not to lick the brush. What else

could we do? It was our work.
Work can cure almost anything.

Little Sister

When her youngest sister was two,
Peg started working at the plant.

Six years later Peg was dead.
The company said anemia.

The company said diphtheria.
She could remember Peg's red hair

and freckles. Their mother rose
at five to walk the mile

to Peg's grave. The factory closed.
White ash seeped into the house,

even with the windows closed.
Peg's little sister dusted every day.

Nothing Sacred

*1937, comedy, running time: 1 hour, 17 minutes, starring Carole
Lombard and Frederic March. French title: La joyeuse suicidée.*

Carole Lombard, as Hazel Flagg, is perky
and adorable as she crosses the sailboat

and nestles into the crook of Wally's arm.
Hazel is Wally's scoop of a lifetime,

the human-interest story that will turn
his newspaper career around. She's

a small-town girl who went to work
at a factory making more money

than she could at a bakery or drug store,
who painted clockfaces to glow in the dark,

and because she pressed her paintbrush
between her lips it's possible she might have

only weeks to live. Papers are flying
off the press. Everyone wants to read

about a beautiful, dying woman.
But there's a catch: dying was just a scare.

Her boozy doctor misdiagnosed her.
I got so that I was seeing radium poisoning

everywhere, he gruffly explains, turning back
to the mirror to shave. As long as she doesn't

confess to Wally her doctor made an error
she'll have a grand time in New York,

where Wally's paper has brought her to give her
a dazzling send-off: take in some shows,

get a key to the city, and be the talk
of the town. What's the harm? The problem is

this is a comedy; Hazel's not dying
like the real girls are. Their doctors say

anemia, pneumonia, hysteria. Their doctors
say radium makes the girls healthier

and prettier. Their doctors are company doctors
while Hazel is the picture of health.

Half-Life

Barely north of where the Fox
and Illinois Rivers join, past a few stretches

of farmland and over the railway line,
Margaret Looney's body lay for forty-nine years

before she was exhumed from St. Columba
and her body driven an hour away

to Argonne National Laboratory
where tests found 19,500 microcuries

of radium in her bones, over a thousand times more
than is safe. Peg worked at Radium Dial

only a few years, so how did Marie Curie
live to sixty-six, after forty years

of touching and refining radium? Curie kept a vial
in her pocket, another in her desk drawer;

at night her workroom glowed with blue-green light
strong enough to read by. A thousand years

after Curie's death even her cookbooks
will stay sealed in lead-lined boxes.

Radium is both constant and capricious.
The time it takes this impish element

to lose half its energy is the span
in which human civilization saw the birth

of Jesus and the first performance
of *A Midsummer Night's Dream,*

a strange tale of creatures roaming a forest
misdirected by a trickster who paints

a love potion on Queen Titania's eyes, filling them
with moonlight: bright and inescapable.

Madame Curie

*1943, biography, drama, romance, running time: 2 hours, 4 minutes,
starring Greer Garson and Walter Pidgeon*

Forty minutes in, Pierre bangs on Marie's bedroom door.
Marie lights a candle and calls out, *Is anything wrong?*

then clutches the covers so as not to expose her nightdress.
Pierre strides into her room, stops a full meter from her bedside,

and suggests they form a perfect scientific partnership,
suggests that like sodium chloride they will be stable

and unchanging until the end of time. She
accepts, and he steps from the room, latching

the door behind him. Four minutes later they're married.
They pose with family and friends, the photographer counting off

ten seconds. We wait, and watch one tableau
after another, a tablecloth barely stirring in the breeze.

At ten a mighty cheer goes up and the Curies bicycle away,
waved off by an adoring and synchronized crowd.

Who could imagine this could happen to a poor, Polish girl
growing up in Russian-dominated Warsaw, not allowed

to continue her studies because she was a woman, who joined
the Flying University, a secret network of classes, smuggling books

under her cloak, bringing what money she had to help gather
a meager library, spine by spine. She studied Polish texts,

Polish history, chemistry, pedagogy. When the Russians
got too close they shifted locations, kaleidoscoping

all over Warsaw to avoid detection. It wasn't until 1891
that Maria Skłodowska was able to go to Paris and earn

degrees in mathematics and physics at the Sorbonne,
and four years later, already his scientific equal, became

Pierre's wife, and this full and perfect rhyme: Marie Curie.
On screen, Greer Garson, fresh off *Goodbye, Mr. Chips*,

Pride and Prejudice, and an Academy Award for *Mrs. Miniver*,
is starring opposite Walter Pidgeon for the fourth time.

She's wide-eyed and unblinking, her posture fierce,
her lips dewy. For all the overly simplified romance and overly

simplified science, it's still impressive that only ten years
after Curie's death from aplastic anemia due to radium poisoning

her life story is being recreated in gauzy black-and-white
by one of the biggest film stars of the 1940s. Those are

some long odds. In the final scene, Greer Garson, white haired,
gives a swelling speech, but the heat has gone out of the film.

It'll be dark soon, all the little lights flickering
in the windows. On the radio, there's another war on.

Mercurochrome

Even though I was too old to need a nap
Grandma made me lie down in the afternoons.

I lay down on the nubbly bedspread for one
full minute. The numbers on the alarm clock

shone greenly, barely aglow, the same pale shade
as dried toothpaste. I tiptoed to the bathroom mirror

and examined my haircut, which looked
like it had been executed with a serrated knife.

Inside Grandma's medicine cabinet was Calamine,
cold cream, Aqua Net, a single lipstick. I stuck

the thermometer under my tongue just to make
the silvery line rise, careful not to fumble it

or I'd have to chase the beads of mercury
along the floor, then I unscrewed the cap

of the Mercurochrome and daubed the wand
across my hand, staining my skin orange-red.

Across the country students marched against the war.
We walked to save on gas. I wasn't suffering

from anything more than boredom.
There was nothing wrong with me.

Celery

A century ago this whole stretch of city
was farmland, thanks in part to the celery rage.

It was touted as the ultimate health food,
a remedy for pain, anxiety, kidney disease,

rheumatism, and constipation; it was so fashionable
that Victorians created cut crystal vases

to display their celery on the table.
Who doesn't get swept up in a desire for a long

and vibrant life, believing the next knock
on the door might bring us just the thing,

a tonic, salve, or tincture to restore and brighten?
You think it could never happen to you

but look at coconut oil, kale, almonds,
margarine, baby formula. Every few years

there's a new craze, something to save us
from dying away, from losing potency.

Life, for all its joy, diminishes our bodies.
Look at this radiant stalk of green.

Why Tooth Fairies Luminesce

My friend has named
her brain tumor Eddie.

He presses against her ocular nerve.
A helmet immobilizes her head

during treatment. Eddie,
bombarded, stops growing,

but the radiation eats
at the lower bone shelf

of her sinus. The infection
is massive and persistent.

She loses teeth. She loses
bone in her cheek.

What's Your Favorite Superpower?

Being able to fly without having to check a bag
would be phenomenal. I move past the bomb-sniffing dogs

and, discreetly so my children don't notice, give
the double bird as I stand with my arms raised,

feet on the yellow shoe prints in the body scan chamber.
Which reminds me: how does x-ray vision work?

How could I peer into someone's pockets but not
scan all the way through to their bones? This seems

an imprecise skill for a superhero, like baking
a loaf of bread over the vent of an erupting volcano.

In fact when scientists discovered x-rays
it wasn't clear what to use them for, all the prints

they made on photographic plates. On December 22, 1895,
Wilhelm Röntgen made a print of his wife's hand

that showed her phalanges and metacarpal bones
and a shadow where she wore her rings, and his wife

responded, *I have seen my death.* How haunting
to see the framework of your body stripped

of the scaffolding of skin and muscle, warmth
and motion. To render the soft tissue invisible,

useless to the eye, is to ignore what makes us
distinctive, recognizable. But what recourse

do we have? I don't want these TSA agents
scrutinizing my bones, the contents of my pockets,

or the underwire in my bra, but I can't shield myself
against their rays or vision. I'm gazed right through,

a pane of glass. Here I am, heading to the terminal.
Here I am, wishing I were as impenetrable as stone.

Radium Again Found in
Allegheny River and Tributaries

Pittsburgh Post-Gazette, 1-19-18

When we walk along the river tonight
with our good black dog, our breath

coming evenly in foggy plumes, all terrors
having receded to tiny pinpricks of light,

we've got one more thing to worry about
before that dog goes crashing off into

the underbrush in search of something
we haven't begun to notice. That's

the nature of radium. It seeps into minewater,
then gets pumped out thirty miles upstream.

How many Superfund sites have we visited
on a family vacation without knowing it? The hotels

were cheap, and the kids just wanted to play Marco
Polo, the pool glowing a deep, unearthly green.

Soap

Either no one's done their homework
or someone's sense of irony runs deep:

this bar of soap stamped *Radium*
stops me in my tracks. In its defense

this museum gift shop has a whole basket
of soaps emblazoned with the names, numbers,

and symbols of all the elements. I wave off
the urge to set out the soaps on the tile floor

in a periodic table, focusing instead
on picking all the radiums out of the basket,

each one with a yellow sticker enthusing
Glows In The Dark! though these soaps charge up

with sunlight or with lamplight; they radiate
what they collect, they don't emit. But here,

in the Museum of Science and Industry,
this soap is exactly where science and industry

intersect, where Marie Curie's radium, painstakingly
distilled from tons of soot-black uranium ore

in the courtyard outside the shed she and Pierre
had been granted as a shabby workspace,

meets the radium-laced paint that rows
and rows and rows of women brushed

onto watchfaces so the numbers would glow,
the radium slowly poisoning their bones,

while factory scientists safeguarded themselves
behind protective shields. Here is science

and industry, and a generous measure of vanity
and foolishness, once the radium craze

took hold a hundred years ago, when elixirs
and creams and tonics infused with radium

promised youth, vitality, brilliance, and shine.
In the end, I buy a single bar. It's hypoallergenic

and is not tested on animals. It is handcrafted
in the USA. Here it sits under my desklamp,

charging. It chants to me *Ra Ra Ra*,
still, after all this time, promising absolution.

NOTES

The Radium Dial Company opened a factory in Ottawa, Illinois in 1922 to be close to Westclox, a major clock manufacturing company in the area. Radium Dial hired young women to paint watch faces with luminescent paint; they were ideally suited for this work, it was believed, because of their fine motor skills and attention to detail. They were explicitly taught to press the paintbrush between their lips to get the brush tip to the finest point possible. In doing so, they ingested radioactive paint, which managers assured them was safe.

Some women started to get sick right away. They had no energy, their bones broke easily, their jaws crumbled when they had teeth pulled. Radium Dial's doctors examined them and claimed they were perfectly healthy, or blamed their illnesses on other causes, like syphilis and diphtheria.

Radium was still relatively young at this point. Marie and Pierre Curie had discovered the existence of radium in 1898; it was Marie Curie who figured out a way to isolate radium from the surrounding material. Together with Henri Becquerel, who discovered radiation, the Curies were awarded the Nobel Prize in Physics in 1903. In 1906 Pierre Curie slipped crossing a rainy street in Paris and was killed by a horse-drawn cart.

During World War I, when the German army was approaching Paris, Marie Curie transported her lab's single gram of radium, which was France's entire research stock, to a safe location in Bordeaux. During the war, Curie persuaded wealthy patrons to donate money and vehicles which she then outfitted into mobile x-ray units; she herself drove one of these contraptions around warzones and helped doctors prepare for surgeries.

For the rest of her life, Marie Curie championed radium and believed in its powers. She respected what it could do but refused to believe radium had any ill effects on her health or anyone else's. While she had nothing to do with it, radium even enjoyed a brief health craze in the early 20th century and was incorporated into all manner of health, wellness, and beauty products.

In the mid-1930s Radium Dial, under pressure from bad publicity stemming from its dying workforce, closed down and reopened soon thereafter and a few blocks away under the name Luminous Processes, which operated until 1978. The factory and the area it contaminated remain an EPA Superfund site.

St. Columba Cemetery in Ottawa, Illinois, where Margaret "Peg" Looney is buried, is the resting place for several women who died as a result of radium poisoning.

Similar factories operated in Orange, NJ and Waterbury, CT. Several "Radium Girls," as they were called, sued for damages; their lawsuits triggered significant advancements in workplace safety and labor law, leading eventually to the Occupational Safety and Health Administration (OSHA) Act which was adopted in 1971.

The half-life of Radium (Ra), or time it takes for its radio-activity to decrease by half, is about 1600 years.

There are many excellent resources available, but some of the most significant to me were:

- *Living With Radiation: The First Hundred Years* (Paul Frame and William Kolb)
- *Madame Curie: A Biography* (Eve Curie, translated by Vincent Sheean).

- *Marie Curie: A Life* (Susan Quinn)
- *Radioactivity : A History of a Mysterious Science* (Marjorie C. Malley)
- *Radioactive: Marie & Pierre Curie: A Tale of Love and Fallout* (Lauren Redniss)
- *Radium Girls: Woman and Industrial Health Reform, 1910-1935* (Claudia Clark)
- *Madame Curie,* a 1943 loosely biographical film directed by Mervyn LeRoy.
- *Marie Curie: The Courage of Knowledge,* a 2016 Polish/French film directed by Marie Noëlle.
- *Nothing Sacred,* a 1937 screwball comedy, directed by William A. Wellman.
- *Radium City,* a 1987 documentary about the Ottawa, IL Radium Girls, directed by Carole Langer.
- *Shining Lives: A Musical.* Book and lyrics by Jessica Thebus; music by Andre Pluess and Amanda Dehnert. Based on the play *These Shining Lives* by Melanie Marnich. The production I saw was directed by Jessica Thebus and performed at Northlight Theater in Skokie, IL in the spring of 2015.

ACKNOWLEDGEMENTS

Many thanks to the editors of the publications in which poems from this book first appeared, sometimes in alternate versions:

Atticus Review: "Mercurochrome"
Colorado Review: "Paradise" and "Work (1923)"
Lake Effect: "Half-Life"
Meridian: "Soap"
Peauxdunque Review: "Madame Curie," "What's Your Favorite Superpower," and "Why Tooth Fairies Luminesce"
Spoon River Poetry Review: "Afterglow," "Work (1922)," and "Work (1934)"

"Marie Curie Addresses Pierre in her Mind While She Inspects the Stove In the Vestibule" is after a poem by Julianna Baggott, "Marie Laurent Pasteur Addresses Louis in Her Mind While She Scalds the Sheets."

Thanks to Mollie Perrot of the Ottawa Historical and Scouting Heritage Museum in Ottawa, IL, who met with me for several hours when I was in the area investigating my family ties to the area. A statue honoring the Radium Girls is about five blocks from the museum.

Thanks to Amy Tudor, who is alive and well. Thanks to Margo Figgins and the Young Writers Workshop of the University of Virginia; a passing conversation there many years ago ignited my curiosity in Marie Curie. Thank you, Margo, for holding that memory with me.

Thanks to my family of poets who support and sustain me, especially Liz Ahl, Joanne Diaz, Lindsay Garbutt, Willie James, Matthew Kelsey, Faisal Mohyuddin, Jeff Oaks, Noah Stetzer, Laura Van Prooyen, Jacob Saenz, Casey Thayer, and my dear *RHINO* crash. Thanks to Ralph Hamilton for his leadership, grace, and insight.

Thanks to Laura Van Prooyen and her generosity in giving these poems a home. I couldn't have imagined a more thoughtful or more attentive press to entrust them to. Thanks to Laura and to Joanne Diaz for helping nurture these poems when they were in their infancy.

Thanks to my students at Chiaravalle Montessori School, who every day show me how to live in the world with curiosity and awe.

And thanks most of all to Allen, Hannah, and Joshua, without whom none of this matters.

ABOUT THE AUTHOR

Ann Hudson is the author of *The Armillary Sphere*, which was selected by Mary Kinzie as the winner of the Hollis Summers Poetry Prize and published by Ohio University Press. A senior editor for *RHINO*, she teaches at a Montessori school in Evanston, Illinois.